Great Explorations in Math and Science (GEMS)

Lawrence Hall of Science,
University of California, Berkeley

SPACE SCIENCE SEQUENCE FOR GRADES 3–5

Unit 3 How Does the Earth Move?

The Space Science Sequence is a collaboration between the
Great Explorations in Math and Science (GEMS) Program
of the Lawrence Hall of Science,
University of California at Berkeley and the
NASA Sun–Earth Connection Education Forum
NASA Kepler Mission Education and Public Outreach
NASA Origins Education Forum/Hubble Space Telescope
NASA Solar System Education Forum
NASA IBEX Mission Education and Public Outreach
Special advisors: Cary Sneider and Timothy Slater
Foreword by Andrew Fraknoi

National Aeronautics and Space Administration
Funding for the GEMS Space Science Sequence was provided by the NASA Forums and Missions listed on the title page.

Great Explorations in Math and Science (GEMS) is an ongoing curriculum development program and growing professional development network. There are more than 70 teacher's guides and handbooks in the GEMS Series, with materials kits available from Carolina Biological. GEMS is a program of the Lawrence Hall of Science, the public science education center of the University of California at Berkeley.

Lawrence Hall of Science
University of California
Berkeley, CA 94720-5200
Director: Elizabeth K. Stage

Project Coordinator: Carolyn Willard
Lead Developers: Kevin Beals, Carolyn Willard
Development Team: Jacqueline Barber, Lauren Brodsky, John Erickson, Alan Gould, Greg Schultz
Principal Editor: Lincoln Bergman
Production Manager: Steven Dunphy
Student Readings: Kevin Beals, Ashley Chase
Assessment Development: Kristin Nagy Catz
Evaluation: Kristin Nagy Catz, Ann Barter
Technology Development: Alana Chan, Nicole Medina, Glenn Motowidlak, Darrell Porcello, Roger Vang

Cover Design: Sherry McAdams, Carolina Biological Supply Co.
Internal Design: Lisa Klofkorn, Carol Bevilaqua, Sarah Kessler
Illustrations: Lisa Haderlie Baker
Copy Editor: Kathy Kaiser

This book is part of the *GEMS Space Science Sequence for Grades 3–5*.
The sequence is printed in five volumes with the following titles:
Introduction, Science Background, Assessment Scoring Guides
Unit 1: *How Big and How Far?*
Unit 2: *Earth's Shape and Gravity*
Unit 3: *How Does the Earth Move?*
Unit 4: *Moon Phases and Eclipses*

Published by Carolina Biological Supply Company. 2700 York Road, Burlington, NC 27215.
Call toll-free 1-800-334-5551.

Printed on recycled paper with soy-based inks.

ISBN 978-0-89278-334-2

UNIT 3

HOW DOES THE EARTH MOVE?

How Does the Earth Move?

SESSION SUMMARIES (4 Sessions)

3.1 Ideas About the Earth and Sun

The primary purpose of the first session is for students to reflect on their ideas about how the Earth moves, setting the stage for the following three sessions. The students fill out a questionnaire and share observations about how the Sun and stars appear to move in the sky.

After a brief lesson (or review) of key concepts about models, students hear about five ancient models that explained the apparent movement of the Sun. They learn that *scientific models* are explanations that are based on evidence.

3.2 Mount Nose

An activity called Mount Nose introduces the students to the current scientific model, now accepted as fact, that explains the Sun's apparent movement in the sky. In this model, a light bulb at the center of the room represents the Sun, and each student's head represents the Earth. Their nose represents a mountain on Earth. Each student slowly turns to simulate the spinning of the Earth. They notice that although the Sun appears to be moving around them, much as it does in the sky, it's actually the Earth itself that is spinning. The model is also used to illustrate why it is daytime on one side of the Earth while it is night on the other side.

Students are then challenged to apply their knowledge by describing the model to someone who thinks the Sun goes around the Earth. In small groups called *evidence circles*, students discuss ways to explain that the Earth's spinning motion causes the apparent motion of the Sun, as well as night and day. Finally, each student writes their explanations on a student sheet which can be used as an assessment before moving on to Session 3.3.

3.3 Spinning Earth

This session reinforces students' understanding of how sunlight, shadow, and the Earth's spin cause the cycle of day and night and the apparent motion of the Sun. This time, students look at the spinning Earth from an outside perspective.

The class first discusses an Apollo image of the Earth in space. Teams of four students then do an activity called *Spinning Globes.* Each team has a globe with colored dots affixed to four locations along the equator. Each student pretends to live in one of these locations. During the activity, they spin the globe and determine whether it is noon, midnight, or some time in-between at each of their locations.

3.4 Earth in Orbit

In this session, the Mount Nose model is used again, but as a demonstration using only one student to model Earth's orbit around the Sun, along with the Earth's spin. Students learn that it takes one year for Earth to orbit the Sun and that there are 365 days in a year. The terms *rotation* and *revolution* are introduced.

Next, the class learns of the Ptolemaic model, which describes the Sun as orbiting around the Earth. Through a reading, they learn of the Copernican model, with the Earth orbiting the Sun, and how Galileo found evidence that supports this theory, which is the model used in modern science.

Students take the *Post–Unit 3 Questionnaire*, which allows them to demonstrate how their ideas about Earth's motion have changed.

Ideas About the Earth and Sun

Overview

The primary purpose of the first session is for students to share their ideas about how the Earth moves, setting the stage for the rest of the unit. The session begins with a questionnaire about positions and movements of the Sun and Earth. After the questionnaires have been collected, the students share observations about how the Sun and stars appear to move in the sky.

After a brief lesson (or review) of key concepts about models, students hear about five ancient models that explain the apparent movement of the Sun. They learn that the Greeks lived at the crossroads of trade routes. Realizing that all the stories they heard couldn't be true, the Greeks attempted to come up with *scientific models* that could be tested, using evidence.

Ideas About the Earth and Sun	Estimated Time
Pre–Unit 3 Questionnaire	20 minutes
Student ideas about Sun and Earth movement	10 minutes
Ancient models explain the Sun's apparent movement	15 minutes
A scientific explanation is based on evidence	15 minutes
TOTAL	**60 minutes**

What You Need

For the class
- ❏ sentence strips for 7 key concepts
- ❏ wide-tip felt pen
- ❏ transparency from the transparency packet or CD–ROM file of *Five Ancient Models of the World*
- ❏ overhead projector or computer with large screen monitor/LCD projector

For each student
- ❏ 1 *Pre–Unit 3 Questionnaire* from the student sheet packet
- ❏ pencil

Getting Ready

1. For each student, make one copy of the *Pre-Unit 3 Questionnaire*.

2. If you will not be using the CD–ROM, make an overhead transparency of *Five Ancient Models of the World*.

Unit Goals

The Earth moves with regular and predictable motion.

The Earth spins (rotates) and orbits the Sun (revolves).

The spinning of the Earth causes the apparent daily movement of the Sun and stars.

Light from the Sun shining on the spinning Earth causes day and night.

The Sun, Earth and Moon form a system.

TEACHER CONSIDERATIONS

TEACHING NOTES

Tips for Leading Good Discussions: Engaging students in thoughtful discussions is a powerful way to enrich their learning. Even the most experienced teacher can use a few reminders about how to lead a good discussion. Listed below are some general strategies to keep in mind while leading a class discussion:

- Ask broad questions (questions which have many possible responses) to encourage participation.
- Use focused questions sparingly (questions which have only one correct response) to recall specific information.
- Use wait time (pause about 3 seconds after asking a question before calling on a student).
- Give non-judgmental responses, even to seemingly outlandish ideas.
- Listen to student responses respectfully, and ask what their evidence is for their explanations.
- Ask other students for alternative opinions or ideas.
- Try to create a safe, non-intimidating environment for discussion.
- Try to call on as many females as males.
- Try to include the whole group in the discussion.
- Offer "safe" questions to shy students.
- Employ hand raising or hand signals to insure whole group involvement.
- Take time to probe what students are thinking.
- Consider your role as a collaborator with the students, trying to figure things out together.
- Encourage students to figure things out for themselves, rather than telling them the answer.

What Some Teachers Said

"We have a 5th grade Science State test. I know my 4th grade students will be confident next year on the space section of the test."

"The children were very engaged. They enjoyed the hands-on lessons and working in groups. The level of challenge was appropriate for most. Even my Russian student (VERY limited English) was able to participate and vocalize the key concepts. Children who need challenging materials also enjoyed the lessons, especially the background information about the early models."

3. Arrange for the appropriate projector format (computer with large screen monitor, LCD projector, or overhead projector) to display images to the class

4. Choose a wall or bulletin board that can serve as a "concept wall" for Unit 3. This is where you will post sentence strips showing the key concepts learned. You'll post the strips in two columns. (Please see the illustration on page 391.) The left column should be titled, *What We Have Learned About Evidence and Models.* The right column should be titled, *What We Have Learned About Space Science.*

5. Write the seven key concepts below on sentence strips and have them ready to post. The first key concept will go in the right column of the concept wall, under *What We Have Learned About Space Science:*

> The Sun and stars appear to move across the sky.

The other six concepts will go on the *What We Have Learned About Evidence and Models* side of the concept wall. *Note: If you have presented Unit 1 or Unit 2, all concepts about evidence and models presented earlier should remain on the concept wall, and all you need to add in this session is the one concept about the apparent movement of the Sun and stars, above.*

> Scientists use models to understand and explain how things work.

> Every model is inaccurate in some way.

> A model can be an explanation in your mind.

> Evidence is information, such as measurements or observations, that is used to help explain things.

> Scientists base their explanations on evidence.

> Scientists question, discuss, and check each other's evidence and explanations.

Unit Goals

The Earth moves with regular and predictable motion.

The Earth spins (rotates) and orbits the Sun (revolves).

The spinning of the Earth causes the apparent daily movement of the Sun and stars.

Light from the Sun shining on the spinning Earth causes day and night.

The Sun, Earth and Moon form a system.

TEACHER CONSIDERATIONS

TEACHING NOTES

Key concepts about models and evidence presented earlier in the sequence: A total of nine key concepts about models and evidence were presented in Units 1 and 2 of the *Space Science Sequence*. If you have presented either unit, leave all those concepts on your concept wall. However, if Unit 3 is your first unit in the sequence, do not present all these concepts now. Because Unit 3 is a short unit, the write-up suggests presenting a total of six key concepts about models and evidence. Just for your information, below are the additional concepts about models that were presented in Units 1 and 2.

Concepts about *models* presented during Units 1 and 2:

Space scientists use models to study things that are very big or far away.
Models help us make and test predictions.
Models can be 3–dimensional or 2–dimensional.

Key Vocabulary

Science and Inquiry Vocabulary

Evidence

Scientific Explanation

Model

Scale Model

Prediction

Scientist

Three–Dimensional (3-D)

Two–Dimensional (2-D)

Space Science Vocabulary

Gravity

Satellite

Orbit

Diameter

Sphere

System

Rotate

Revolve

PRE-UNIT 3 QUESTIONNAIRE, PAGE 1

Session 3.1 Student Sheet   Name _____

Pre–Unit 3 Questionnaire, Page 1

Note: Pictures are not to scale.

1. Where is it night on Earth in this picture? Color in the areas where it is night.

2. Why does it look like the Sun is moving down in the sky at sunset? Circle the best answer.

A. The Sun goes around the Earth.
B. The Earth is spinning.
C. The Earth goes around the Sun.
D. The Sun is spinning.

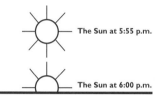

3. Which drawing best shows how the Sun and Earth move?

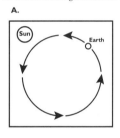 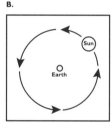 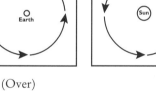

(Over)

Pre–Unit 3 Questionnaire

1. Questionnaire about Earth's movement. Tell the class that they will be studying how the Earth moves. Let them know that you want to find out their current ideas by having them each fill out a questionnaire.

2. Write your own ideas. Tell students that the questionnaire's purpose is to find out what each of them is thinking, so it is not helpful if they share ideas with one another at this point. Tell them that if they aren't sure of an answer on the questionnaire, they should just put down their ideas. Tell them that this questionnaire will not affect their grade. Hand out a copy of the questionnaire to each student. Have them write their names at the top of their papers and answer the questions.

3. Early finishers. You may want to suggest independent work for students who finish early. Allow time for all students to finish. Collect the questionnaires.

Students' Ideas About the Movements of the Sun and Stars

1. **Where in the sky have students seen the Sun?** Guide the class to a general agreement about the Sun's positions in the sky each day by asking the following questions:

- Where in the sky have you seen the Sun?
- Where was the Sun yesterday at sunset? [West.]
- Point in the direction that you think the Sun was at sunset. [West.]
- Point in the direction that you think the Sun will be at noontime today. [High in the sky.]
- Point in the direction that you think the Sun will be at sunrise. [East.]

Some students may be aware that in the northern hemisphere, the Sun rises and sets farther to the south in October through February, and farther to the north in April through August.

TEACHER CONSIDERATIONS

TEACHING NOTES

Drawings not to scale: Be sure that students understand that the relative sizes of the Earth and Sun are NOT TO SCALE in the drawings on the questionnaire. Also, the size of the person on Earth in question #4 is greatly enlarged. Students should focus on the position of objects and their movements, not their sizes.

QUESTIONNAIRE CONNECTION

We have included *Questionnaire Connection* notes in later sessions of the unit to help you highlight when an activity or discussion relates to specific questions on the questionnaire. It is important to look over your students' questionnaires before getting too far into the unit, so you can identify areas that may need more or less instruction, and any major misconceptions that students may have.

Scoring Guide for the *Pre–Unit 3 Questionnaire*: The questionnaire can be scored using the scoring guide on page 82.

Different Order of Questions on the *Pre and Post Questionnaires*: The *Post–Unit 3 Questionnaire* in Session 3.4 is identical in content, but the placement of some questions and the order of the possible answers are different than on the *Pre–Unit 3 Questionnaire*. This is to encourage students to answer thoughtfully, rather than simply remembering the order of the answers.

Name:_____

Post-unit 3 Questionnaire continued

3. Here is an enlarged person standing on the Earth. It is noon for this person. Which direction is the Sun? Circle A, B, C, or D.

4. Why does it look like the Sun is moving down in the sky at sunset? Circle the best answer.

A. The Earth goes around the Sun.
B. The Sun goes around the Earth.
C. The Sun is spinning.
D. The Earth is spinning.

The Sun at 5:55 p.m.

The Sun at 6:00 p.m.

5. Where is it night on Earth in this picture? Color in the areas where it is night.

Earth

Sunlight

What Some Teachers Said

"I thought they would do better than they did on the pre test. It was an eye opener."

"I was surprised at some of the ideas my students had at the beginning of this unit. I was able to see growth in their understanding on the post questionnaire."

2. How do the stars move in the sky? Ask students if they have noticed that the stars change position in the sky. If so, ask them to describe the motion of the stars. [Depending on their access to the night sky, some students may have observed constellations of stars rising in the east and setting in the west.] *It is quite common for students not to have observed the apparent movement of the stars.*

3. Introduce the concept wall. Post the key concept and read it aloud with the students. Say that people have observed the Sun and stars for thousands of years and watched them seem to move across the sky. Tell them that people have come up with different explanations for what they observed.

> The Sun and stars appear to move across the sky.

4. How does the Sun get back to the east? Ask, "After the Sun sets in the west, how does it get back to the east before it rises the next morning?" Encourage several answers.

Ancient Models to Explain the Sun's Apparent Movement Across the Sky

1. What is a model? If you didn't do so in Unit 1 or 2 of the Sequence, define *models.* (Please see page 359.) Tell students that they will be using models during the coming activities. Point out that even though all models have their limitations and have some inaccuracies, a good model is accurate in one way or in enough ways to make it useful to scientists.

2. Add models concepts to concept wall. Post the three concepts below about models. Read them with students, and clarify any terms that they may not have encountered earlier. Emphasize that a model can be a physical object, but it can also simply be a person's explanation for something they have observed.

> Scientists use models to understand and explain how things work.
>
> Every model is inaccurate in some way.
>
> A model can be an explanation in your mind.

Unit Goals

The Earth moves with regular and predictable motion.

The Earth spins (rotates) and orbits the Sun (revolves).

The spinning of the Earth causes the apparent daily movement of the Sun and stars.

Light from the Sun shining on the spinning Earth causes day and night.

The Sun, Earth and Moon form a system.

TEACHER CONSIDERATIONS

TEACHING NOTES

More on Models: If this is the first time that your students have encountered the concept, you may want to take some extra time to explain models and how scientists use them, as is done in Unit 1, Session 1.4, of the *Space Science Sequence*:

1. **Introduce models.** Hold up a scale model car (or any model), and say that a model is something that shows or explains what the real thing is like.

2. **All models are different from the real thing in at least one way.** Emphasize that good models are like the real thing, but no model is exactly the same as the real thing. Ask, "What are some ways the model car is not exactly the same as a real car?" [It's smaller, has no motor, the doors don't open, the tires are metal, it doesn't have gas in it, it has no lights, it can't move under its own power, and so on.]

3. **Define scale models.** Tell them that although it's much smaller than a real car, this model looks like a car because someone measured every part and made them all smaller by the same amount. It is a scale model of a real car.

4. **Scientists often use models.** Say that scientists use models to explain things they have observed, show how they think things in the natural world work, make predictions, or learn more about things they can't look at directly. In space science, models are used a lot, because the things being studied are so far away and large.

Key Vocabulary

Science and Inquiry Vocabulary

Evidence

Scientific Explanation

Model

Scale Model

Prediction

Scientist

Three–Dimensional (3-D)

Two–Dimensional (2-D)

Space Science Vocabulary

Force

Mass

Gravity

Satellite

Orbit

Diameter

Sphere

System

Rotate

Revolve

The Earth moves
with regular and
predictable motion.

The Earth spins
(rotates) and orbits
the Sun (revolves).

The spinning of
the Earth causes
the apparent daily
movement of the
Sun and stars.

Light from the
Sun shining on the
spinning Earth causes
day and night.

The Sun, Earth and
Moon form a system.

3. Describe five ancient models of the world. Tell students that people in ancient times proposed different models for how the Sun moves, including how it is able to get back to the east again every morning. Tell students that if they had gone to school 3,000 years ago, they might have been taught that one of these models is the way to explain observations of the Sun.

4. Show the *Five Ancient Models of the World* transparency, and briefly describe the five models, as summarized below. Emphasize that these explanations are all from thousands of years ago. Also tell students that in each of the places you mention, there were many different stories about the world.

Ancient India: The Earth is a flat, circular disk, surrounded by ocean. In the center of the world is a great mountain. Every evening, the Sun goes behind the western side of the mountain. It travels behind the mountain at night, and comes out on the eastern side in the morning.

Ancient Egypt: The Earth is flat. The sky is like a flat plate, supported at four places by mountains. The Sun is carried across the sky in a boat, from east to west. At night, the Sun is carried back to the east on a river through the underworld.

Ancient China: The sky is a round dome, surrounding a flat, square-shaped Earth. The ocean goes all around the Earth. The Sun travels in a big, tilted circle. At night, the Sun is not under the Earth. Instead, it is on the side of the Earth.

Ancient Greece: The Earth floats in the ocean like a cork in water. The Sun god drives his chariot across the sky each day, resting his horses under the Earth at night.

Ancient Mexico: The Earth and sky are two halves of a monster. There is water to the east and west. The sky is held up by five trees. The Sun rises in the east, and travels west. It is swallowed up by the monster as it sets. It travels through the underworld and rises again the next day.

TEACHER CONSIDERATIONS

TEACHING NOTES

Many other possible models: There are many other possible models from world cultures that could be used as examples in this session. The GEMS guide *Investigating Artifacts* has a more extensive list of stories that explain natural phenomena, including the movements of the Sun and the Moon.

PROVIDING MORE EXPERIENCE

Students Create "Ancient Models." Before exploring the scientific model in this unit, you may want to give your students more experience thinking about ancient models for the apparent movement of the Sun. The following activity gives students a chance to examine their own observations of and explanations for the Sun's movement:

FIVE ANCIENT MODELS OF THE WORLD

ANCIENT INDIA

ANCIENT EGYPT

ANCIENT CHINA

ANCIENT GREECE

ANCIENT MEXICO

1. Have your students imagine that they lived several thousand years ago, on the site where they live now. Challenge them to invent a model of the world to explain how the Sun gets from the western part of the sky back to the east during the night. The model can be a flat Earth or any other shape that might explain the observations.

2. Organize the students into pairs, and hand out scratch paper so that they can sketch their ideas. (Note: This project can also be done by individuals or small groups.)

3. Give each pair of students a large sheet of white paper. Explain that they should draw their ideas of the world so the drawing can be seen by the entire class. Suggest that they label their drawings. If time allows, have them explain their "ancient models" to the class.

Greece, Egypt, India, China, and many other nations and cultures contributed to advances in astronomy. Ideas were exchanged, debated, and built on each other as modern science developed. Many indigenous cultures, such as those of South, Central, and North America, also developed a high degree of astronomical expertise.

Unit Goals

The Earth moves with regular and predictable motion.

The Earth spins (rotates) and orbits the Sun (revolves).

The spinning of the Earth causes the apparent daily movement of the Sun and stars.

Light from the Sun shining on the spinning Earth causes day and night.

The Sun, Earth and Moon form a system.

5. **Discuss the explanations.** Point out that many ancient models describe the Earth as flat, although we know today that it is *spherical*, or shaped like a ball. Also mention that people from each of these countries came up with many very different explanations for the daily movement of the Sun, and these are just samples.

6. **How might each model reflect the surroundings of the people who created them?** Ask students for some ideas about how the places where ancient peoples lived might have affected their models. You may want to share some information that might explain how the models reflected their surroundings or items they used:

- People from India could not get beyond the steep Himalaya mountains.

- Egyptians made flat metal plates and lived in a river valley.

- The Chinese made both round and square ceramic and metal bowls.

- Many ancient Greeks lived on islands, surrounded by the sea.

- Mexico has ocean on both sides. Many cultures envisioned the Sun as being "swallowed" each time it sets.

A Scientific Explanation is Based on Evidence

1. **A model can be an explanation in your mind.** Review the key concept that a model can be any picture or explanation that someone has in their mind that explains something they have observed.

2. **Some ancient thinkers heard about different models, and realized that they couldn't all be true.** Tell the class that, as different places became centers of trade routes, people from many countries and cultures met and exchanged ideas about their models of the Earth and sky. Some early thinkers heard these different ideas, knew they couldn't *all* be true, and began to develop more scientific ways of thinking about the natural world and deciding if a model was accurate.

TEACHER CONSIDERATIONS

ASSESSMENT OPPORTUNITY

Critical Juncture—Understanding the scale and shape of the Earth:
It is necessary for students to understand the spherical shape of the Earth before moving on to later sessions in Unit 3, in which they will explore the scientific model for day and night and the apparent movement of the Sun and stars.

Many students harbor misconceptions about the shape and scale of the Earth. Young children may construct private models to reconcile what they are told about the round Earth with the evidence of their eyes. For example, because the Earth looks flat, they may envision the Earth as "round," but like a plate. Students who do not have a grasp of the large scale of the Earth compared with people may have trouble accepting the spherical model.

Give special attention to the discussion of ancient models and of students' own models in this session. If students need it, take some extra time to present the activities in *Providing More Experience,* below.

PROVIDING MORE EXPERIENCE

The spherical shape of the Earth is explored fully in the first two class sessions of Unit 2, *Earth's Shape and Gravity.* If your students have not experienced Unit 2 or learned about the Earth's shape in another way, consider inserting those two sessions now, before going on to the rest of Unit 3. If time is short, use as many of the activities in those sessions as possible to provide your students with evidence of the Earth's spherical shape. Two possibilities are the virtual *Trip Around the Earth* in Session 2.2, or the student reading in which Yuri Gagarin describes the Earth's shape. Having students use globes and discuss questions #1 and #3 of the *Pre–Unit 2 Questionnaire* would also be very helpful.

If your students have not experienced Unit 1, they may have some confusion about the Earth's shape related to the issue of scale. To help them understand that they see only a tiny part of the globe, and that is why the ground may look flat, you might use the optional *Insect on the Globe* or *Nose Against a Building* examples on page 249 in Unit 2, Session 2.

What One Teacher Said

"My students were very interested in the description of the models used by ancient cultures. They all decided they... would demand evidence. (Quite a change for this class from the beginning of this Sequence)."

Key Vocabulary

Science and Inquiry Vocabulary

Evidence

Scientific Explanation

Model

Scale Model

Prediction

Scientist

Three–Dimensional (3-D)

Two–Dimensional (2-D)

Space Science Vocabulary

Force

Mass

Gravity

Satellite

Orbit

Diameter

Sphere

System

Rotate

Revolve

3. **A search for evidence.** In more recent times, scientists accept a model only if they can test the model and find *evidence* to support it. Explain what evidence is, and post the following key concepts on the concept wall:

Evidence is information, such as measurements or observations, that is used to help explain things.
Scientists base their explanations on evidence.
Scientists question, discuss, and check each other's evidence and explanations.

4. **A scientific explanation.** A *scientific explanation* or model is based on evidence. By using evidence to explain how the Sun seems to move in the sky and other things they observed, some early thinkers concluded that the Earth is ball-shaped. Tell the students that in the next class session they will investigate the scientific model for why the Sun seems to move the way it does.

Unit Goals

The Earth moves with regular and predictable motion.

The Earth spins (rotates) and orbits the Sun (revolves).

The spinning of the Earth causes the apparent daily movement of the Sun and stars.

Light from the Sun shining on the spinning Earth causes day and night.

The Sun, Earth and Moon form a system.

TEACHER CONSIDERATIONS

OPTIONAL PROMPTS FOR WRITING OR DISCUSSION

You may want to have students use one or both of the prompts below for science journal writing at the end of this session or as homework. These could also be used for a discussion or during a final student sharing circle.

• What are some questions you have about what causes night and day?

• When it is night on Earth, where is the Sun?

What One Teacher Said

"They're really beginning to understand and look for how models help scientists change the way they think. Also that models are in our heads and we can refer to those models as well."

Key Vocabulary

Science and Inquiry Vocabulary

Evidence

Scientific Explanation

Model

Scale Model

Prediction

Scientist

Three–Dimensional (3-D)

Two–Dimensional (2-D)

Space Science Vocabulary

Force

Mass

Gravity

Satellite

Orbit

Diameter

Sphere

System

Rotate

Revolve

Overview

This activity introduces students to the current scientific model, now accepted as fact, that explains the Sun's apparent movement in the sky. In this model, a light bulb at the center of the room represents the Sun, and each student's head represents the Earth. Each student's nose represents a mountain on Earth. Each student slowly turns to simulate the spinning of the Earth. As the Earth spins, the student notes the position of the Sun in the sky and the time of day for a person standing on Mount Nose. They notice that although the Sun appears to be moving around them, much as it does in the sky, it's actually the Earth itself that is spinning. The model is also used to illustrate why it is daytime on one side of the Earth while it is night on the other side.

Students are then challenged to apply their knowledge by describing the model to someone who thinks that the Sun goes around the Earth. In small groups called *evidence circles*, students discuss ways to explain that the Earth's spinning motion causes the apparent motion of the Sun, as well as night and day. Finally, each student writes their explanations on a student sheet, which can be used as an assessment before moving on to Session 3.3.

Mount Nose	Estimated Time
The scientific model: Mount Nose and discussion	25 minutes
Evidence circles: Discussion and writing	35 minutes
TOTAL	**60 minutes**

What You Need

For the class
- ❑ 1 light bulb base or lamp with no shade
- ❑ 1 40-watt (or greater) light bulb
- ❑ 1 extension cord
- ❑ duct tape or masking tape to tape down cord
- ❑ wide–tip felt pen
- ❑ sentence strips for four key concepts

For each student
- ❑ *Evidence Circles: How Does the Earth Move?* student sheet, from the student sheet packet

Unit Goals

The Earth moves with regular and predictable motion.

The Earth spins (rotates) and orbits the Sun (revolves).

The spinning of the Earth causes the apparent daily movement of the Sun and stars.

Light from the Sun shining on the spinning Earth causes day and night.

The Sun, Earth and Moon form a system.

TEACHER CONSIDERATIONS

SCIENCE NOTES

If the Earth is spinning, why can't we feel the motion? A question that students often ask is, "Why can't we feel the spin of the Earth?" We don't feel this spin because the atmosphere and everything else on the ground is rotating with us, and because the spinning motion is extremely smooth. Think of sitting in a car and reading a book or playing a game. If the car were moving very smoothly, with no stops or bumps, if you didn't look out the window, and if the windows were closed, then you really wouldn't know that you were moving. In a similar way, we don't feel or hear any "whoosh" related to our approximately 1,000 miles per hour spinning motion in space because our atmosphere is moving right along with us. And we don't get "thrown off" the Earth by this spinning motion because the Earth's gravity holds us down on the surface.

How fast is the Earth spinning? Students may also ask how fast the Earth is spinning. At the equator, the Earth's spin speed is about 1,670 kilometers per hour, or about 1,040 miles per hour. Away from the equator, the speed is slower because the radius of the spin circle (the distance from Earth's surface to its rotation axis) is less, thus so is the circumference of the rotation circle. As you move close to either the North Pole or South Pole, in fact, the speed of rotation approaches zero. The speed quoted above is derived from the Earth's circumference around the equator divided by the number of hours in one rotation (24 hours = one day).

Key Vocabulary

Science and Inquiry Vocabulary

Evidence

Scientific Explanation

Model

Scale Model

Prediction

Scientist

Three–Dimensional (3-D)

Two–Dimensional (2-D)

Space Science Vocabulary

Force

Mass

Gravity

Satellite

Orbit

Diameter

Sphere

System

Rotate

Revolve

Getting Ready

1. Decide where you will gather your students in a circle around the light bulb for the Mount Nose activity. You may want to have students move some desks aside temporarily during the session. Or you can plan to have students form a circle around some or all of the desks. It is not necessary to darken the room for this activity, but it is desirable.

2. Use the extension cord to plug in the lamp and set it up in the middle of the area where students will form the circle. It's best if the light bulb is at roughly eye level for most students. Tape the cord down to the floor for safety.

3. Make a copy for each student of the *Evidence Circles: How Does the Earth Move?* student sheet.

4. Prepare sentence strips for the following four space science concepts introduced during this session. Have them ready to post during the session under the *What We Have Learned About Space Science* side of the concept wall:

The Earth spins.
The spinning of the Earth makes it look as though the Sun and stars are moving.
It takes the Earth 24 hours to spin once.
The Earth spinning in sunlight causes day and night.

Mount Nose: A Scientific Model to Explain the Apparent Movement of the Sun

Note: With the Mount Nose model, you begin by giving students a chance to witness the Sun rising and setting from a viewpoint on Earth. Once students understand that it is the spinning Earth that causes the apparent movement of the Sun, you use the model to explain global night and day.

TEACHER CONSIDERATIONS

TEACHING NOTES

Why wait to introduce the terms *rotation* and *revolution?* Many adults have trouble remembering the difference between *rotation* and *revolution,* as used to describe the Earth's movements. This is probably because these words are used almost interchangeably in familiar usage and in other contexts. In astronomy, however, the terminology is very specific. Rotation is used to describe a body's spin. (The Earth rotates on its axis.) Revolution is used to describe a body's orbit around something else. (The Earth revolves around the Sun.)

We have chosen to use the word *spin* in this activity, rather than *rotation*, and *orbit* in Session 3.4, rather than *revolution*. We have found it easier initially for students to grasp the concepts using these more readily distinguished terms. We have also chosen to use these terms on the questionnaires, to prevent students who understand the concept from becoming confused by the terms and marking the questionnaires inaccurately. Although it is important for students to learn the terms rotation and revolution, especially if those words are used in your state standards, we recommend beginning with spin and orbit, then changing to rotation and revolution when you think it is appropriate for your students.

What Some Teachers Said

"My students loved Mount Nose. I could really see an impact on the students' understanding after we finished these activities. They were all eager to demonstrate and describe what they learned in this lesson. Most of my students like to share, but this is the first time this year that I had all my students excited about learning more."

" Such a simple way to explain a difficult concept. I'm sure my students will remember this lesson and therefore the concept."

Unit Goals

The Earth moves with regular and predictable motion.

The Earth spins (rotates) and orbits the Sun (revolves).

The spinning of the Earth causes the apparent daily movement of the Sun and stars.

Light from the Sun shining on the spinning Earth causes day and night.

The Sun, Earth and Moon form a system.

1. **Scientific model to explain why the Sun rises and sets.** Let students know that they will now see a scientific model for what causes us to see the Sun rise, move, and set. This model has been around for a long time, and because it has so much evidence supporting it and no evidence that does not support it, it is the model accepted by scientists today.

2. **Students form a circle.** Have the class stand in a circle around the light bulb. Turn on the light bulb and turn off the classroom lights. Explain that in this model, each of their heads represents the Earth. The light in the center represents the Sun.

3. **Inaccuracies in the model.** Ask, "What is inaccurate about this model?" [The Sun is actually much bigger than the Earth, and the Sun and Earth are much farther apart than in this model.] Say that even though this model's scale is not accurate, it is useful to explain why the Sun rises and sets.

4. **Introduce Mount Nose.** Ask students to imagine that each of their noses is a mountain called *Mount Nose,* and that a person lives on the tip. Point out that this is another inaccuracy of scale in the model, because there is no mountain that big on Earth.

5. **What time is it on Mount Nose?** With the students facing the light bulb, ask, "For the person standing on your Mount Nose, where in the sky is the Sun?" [High in the sky, overhead.] Ask, "What time of day do you think it is for the person on Mount Nose?" [Around noon.]

6. **The Earth spins.** Have everyone turn to their left, and stop when their right ears are facing the Sun. Ask, "For the person on Mount Nose, where in the sky does the Sun seem to be?" [Near the horizon, low in the sky.] Ask, "What time of day is it for the person?" [Sunset.] Ask them to watch as the Sun appears to "set" the next time they turn.

7. **Backs to the light.** Have the students make another quarter turn, stopping when their backs are to the light bulb. Ask, "What time is it now for the person on Mount Nose?" [Around midnight.]

8. **Left ears toward the Sun.** Have the students make another quarter turn, so that their left ears face the Sun. Ask, "Where is the Sun for the person on Mount Nose?" [Low in the sky, just "coming up."] "What time is it on Mount Nose?" [Sunrise.]

9. **Do one or two more spins.** Have students slowly do a few more complete turns on their own, watching as the Sun appears to rise, move across the sky, and set again.

TEACHER CONSIDERATIONS

PROVIDING MORE EXPERIENCE

A tiny figure standing on the summit of Mount Nose: To help students with spatial reasoning during the Mount Nose activity, one teacher gave each of her students a tiny plastic human figure. Each student held a tiny human figure in a standing position on the tip of his or her nose while doing the Mount Nose activity. This helped students understand where the Sun would be in the sky in relation to the person (for example, that the Sun would be overhead at noon). The teacher was careful to point out to students that the scale of the tiny plastic figure was even more inaccurate than the scale of Mount Nose itself.

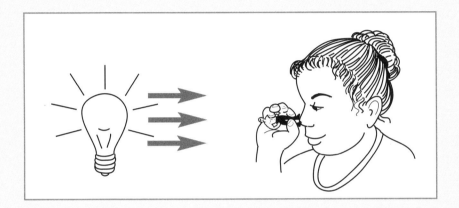

Key Vocabulary

Science and Inquiry Vocabulary

Evidence

Scientific Explanation

Model

Scale Model

Prediction

Scientist

Three–Dimensional (3-D)

Two–Dimensional (2-D)

Space Science Vocabulary

Force

Mass

Gravity

Satellite

Orbit

Diameter

Sphere

System

Rotate

Revolve

10. **The Sun was stationary.** Ask, "Why did the Sun look as though it were moving to the person on Mount Nose?" [Because the Earth was spinning.] Ask, "Did the Sun really move?" Emphasize that the Sun did not move. Scientists today use a model like the one they have just made to explain why the Sun seems to move in the sky.

How the Mount Nose Model Explains Night and Day

1. **One spin is 24 hours.** Keep students standing around the light bulb, and tell them that this model can also explain night and day. Ask if anyone knows how long it takes the real Earth to spin one time around. [24 hours.]

2. **Where is it night?** While they are facing the light, say, "Point to a part of your head where it is dark." [The back of the head.] Tell them that this is where it is night. Make sure that everyone notices that while it is daytime on Mount Nose, it is nighttime on the other side of the Earth.

3. **Where is it day?** Have them turn to the left until their backs are to the light, and say, "Point to the part of your head where it is daytime now." [The back of the head, because it is now facing the Sun.] "What time is it on Mount Nose?" [Midnight.]

4. **Watch night and day on one student.** Have the class watch as one student spins slowly, observing how the shadow darkens different parts of the head as it moves. Have everyone watch the student's Mount Nose and call out "day" and "night" as the student spins.

5. **The stars seem to move when the Earth spins.** Mention that at night, as the Earth spins, the stars seem to move across the sky, just as the Sun does during the day. Emphasize that the stars only *seem* to move. Turn on the classroom lights, turn off the light bulb, and have the class return to their seats.

Unit Goals

The Earth moves with regular and predictable motion.

The Earth spins (rotates) and orbits the Sun (revolves).

The spinning of the Earth causes the apparent daily movement of the Sun and stars.

Light from the Sun shining on the spinning Earth causes day and night.

The Sun, Earth and Moon form a system.

TEACHER CONSIDERATIONS

SCIENCE NOTES

The Sun spins, too: Although the light bulb "Sun" in this model is completely stationary, the real Sun is not. The Sun spins, too. Galileo was able to observe sunspots move across the surface of the Sun, and figured out how fast the Sun spins by keeping track of how sunspots move. It takes 27 Earth days for the Sun to spin around one time at the Sun's equator. But because the Sun is not solid, but gas, the North and South poles of the Sun take longer to spin around. They take almost 35 Earth days.

QUESTIONNAIRE CONNECTION

The Mount Nose activity relates to most of the questions on the *Pre-Unit 3 Questionnaire*. You might want to revisit several questions (#1, #2, and #4 especially, and perhaps #5) either now, or after Session 3.3.

TEACHING NOTES

More on Shadows: Many students think of a shadow as the "Peter Pan" shadow that is cast onto other surfaces by an object that is blocking the light. They may have difficulty understanding that night on Earth is caused by the Earth's own shadow on itself. In Unit 4, Session 4.1, a *Shadow Play* activity addresses common student misconceptions about the Earth's shadow.

It's Following Us: A teacher told us that a few of her students remained convinced that the Sun is moving "because it follows us when we are in the car." This is an interesting *optical illusion* that is also experienced for the Moon. If some of your students suggest this, and you have not done Unit 1 of this sequence (on scale, including distance, size, and apparent size), you might want to consider presenting sessions from it.

Key Vocabulary

Science and Inquiry Vocabulary

Evidence

Scientific Explanation

Model

Scale Model

Prediction

Scientist

Three–Dimensional (3-D)

Two–Dimensional (2-D)

Space Science Vocabulary

Force

Mass

Gravity

Satellite

Orbit

Diameter

Sphere

System

Rotate

Revolve

6. Post key concepts. Read and review each of the following key concepts as you add them to the concept wall:

The Earth spins.
The spinning of the Earth makes it look as though the Sun and stars are moving.
It takes the Earth 24 hours to spin once.
The Earth spinning in sunlight causes day and night.

Evidence Circles: Explaining the Scientific Model

1. A dialogue with a person from ancient times. Ask the class to pretend that they could go back in time 3,000 years and talk to a person in ancient times. Pretend that the person says, "Each day, I see the Sun rise, go across the sky, and set. I think that this is because the Sun goes around the Earth every day."

2. It's easy to see why ancient people might have thought this. Acknowledge that it is understandable that someone would think that the Sun goes around the Earth, but that this is not the scientific model. We know now that the Sun does not go around the Earth.

3. Evidence circles. Say that they will be working in small teams called *evidence circles* to explain to the imaginary person from the past why the Sun seems to move the way it does. They will use their experience with the Mount Nose model to explain why the Sun seems to rise, move, and set each day.

4. Divide the class into teams of about four students. Put the students in groups of four, and have students in each team number off from one to four.

Unit Goals

The Earth moves with regular and predictable motion.

The Earth spins (rotates) and orbits the Sun (revolves).

The spinning of the Earth causes the apparent daily movement of the Sun and stars.

Light from the Sun shining on the spinning Earth causes day and night.

The Sun, Earth and Moon form a system.

TEACHER CONSIDERATIONS

ASSESSMENT OPPORTUNITY

Embedded Assessment: The main goal of the evidence circle discussion and writing assignment is to deepen students' understanding of the scientific model for Earth's movement, presented in the Mount Nose activity. (Students will gather more evidence about how the Earth moves in Sessions 3.3 and 3.4.)

As in other units of the *Space Science Sequence,* evidence circles are one way that students act as scientists. Students' written responses to the questions provide an opportunity to evaluate each individual students' ability to write a coherent and compelling argument based on evidence from key concepts presented in Sessions 3.1 and 3.2. An "understanding science concepts" rubric specifically created for this assessment is included on page 377. The assessment can also be scored using the rubrics on page 66.

Session 3.2 Student Sheet   Name _____
Evidence Circles: How Does the Earth Move?

A person 3,000 years ago might have said, "Each day, I see the Sun rise, go across the sky, and set. I think that this is because the Sun goes around the Earth every day."

Pretend that you can talk with a person from 3,000 years ago who thinks that the Sun goes around the Earth. Try to help the person understand the scientific model. Write the answers to the questions below.

1. Why does the Sun seem to move across the sky? _____
_____
_____
_____

2. What causes sunrise? _____
_____
_____
_____

3. What causes sunset? _____
_____
_____
_____

4. What causes night and day? _____
_____
_____
_____

OPTIONAL PROMPT FOR WRITING

You may want to have students use the prompt below in their science journals at the end of this session or as homework:
Pretend that you are looking down from the ceiling on the Mount Nose model. Draw the model, including one person's head, the light bulb, and a person standing on Mount Nose. Draw the model showing one of these times for the person on Mount Nose: sunset, sunrise, or midnight.

5. **Procedure for discussion and writing.** Pass out the *Evidence Circles: How Does the Earth Move?* student sheet to each student. Explain the procedure for evidence circles:

> **a. The first student in an evidence circle reads and answers question #1.** Student #1 will read the first question and try to answer it out loud, as if they were talking to someone who doesn't understand how the Earth moves.
>
> **b. Each person has a turn to speak.** Next, everyone in their team will have a turn to add anything helpful. The team should discuss all the evidence and arguments they can think of to help the person from 3,000 years ago understand how the Earth moves.
>
> **c. Everyone writes.** After everyone in an evidence circle has had a chance to talk, each person will write down on the student sheet their best explanation for question #1.
>
> **d. Same procedure for the other three questions.** Explain that students #2, 3, and 4 will each read their questions, the team will discuss them the same way, and everyone will write down their best explanations.

6. **Use the evidence from the Mount Nose model.** Tell students to use their experience with the Mount Nose model as they try to help the person from ancient times understand why the Sun seems to move. Mention that they can also refer to the key concepts (on the concept wall) about how the Earth moves. Have them begin.

7. **Circulate during discussions.** As students work, circulate to teams and make sure that everyone has a turn to speak. Remind students to think about the Mount Nose model as they put their explanations into words.

8. **Class discussion.** When students have finished, ask one or two volunteers to read aloud their responses to question #1, and ask if others have anything to add. Do this for each statement. If students seem to have trouble addressing any of the questions, spend a little time discussing them as a class.

Unit Goals

The Earth moves with regular and predictable motion.

The Earth spins (rotates) and orbits the Sun (revolves).

The spinning of the Earth causes the apparent daily movement of the Sun and stars.

Light from the Sun shining on the spinning Earth causes day and night.

The Sun, Earth and Moon form a system.

TEACHER CONSIDERATIONS

RUBRIC FOR EMBEDDED ASSESSMENT:

Evidence Circles: How Does the Earth Move?

Student progress and understanding can be assessed with the specific *Understanding Science Concepts* rubric below or with the general rubrics provided on page 66.

	Understanding Science Concepts The key science concepts for this assessment are the following: 1. The Earth spins (rotates). 2. The spinning of the Earth causes the apparent daily movement of the Sun. 3. Light from the Sun shining on the spinning Earth causes day and night.
4	The student demonstrates a complete understanding of all of the key science concepts and uses scientific evidence to support the written explanation. The student needs to mention that the spinning of the Earth causes the apparent movement of the Sun across the sky each day, sunrise, sunset, or day and night. Evidence from the Mount Nose activity should be used to support their argument.
3	The student demonstrates a partial understanding of the key science concepts. Although understanding is demonstrated, the student does not tie all of these concepts together in the explanation and does not support the explanation with evidence from the Mount Nose class activity. The "evidence" may be more generally experiential or rely on statements of presumed fact, rather than being drawn directly from classroom science experiences.
2	The student demonstrates an insufficient understanding of the science concepts. The student demonstrates an understanding of one of the key concepts, but does not demonstrate an understanding of all of the concepts and does not use evidence from class to support the explanation.

1	The information is inaccurate.
	 Some possible inaccuracies are
	   a. The Sun appears to move across the sky, rise, set, or day and night occur because the Sun orbits around the Earth.
	   b. The Sun appears to move across the sky, rise, set, or day and night occur because Earth orbits around the Sun.
	   c. The Sun appears to move across the sky, rise, set, or day and night occur because the sky is a round dome and the Sun is on the other side of the Earth at night.
	   d. The Sun appears to move across the sky, rise, set, or day and night occur because the Sun is driven across the sky by the Sun god in his chariot.

0	The response is irrelevant or off topic.
n/a	The student has no opportunity to respond and has left the question blank.

Key Vocabulary

Science and Inquiry Vocabulary

Evidence

Scientific Explanation

Model

Scale Model

Prediction

Scientist

Three–Dimensional (3-D)

Two–Dimensional (2-D)

Space Science Vocabulary

Force

Mass

Gravity

Satellite

Orbit

Diameter

Sphere

System

Rotate

Revolve

Overview

This session reinforces the concept that Earth's spin causes day and night which was presented in Session 3.2. In this session models are used to look at the spinning Earth from an outside perspective. The class first looks at an *Apollo 11* image of the Earth in space, as well as a spinning globe, and studies how sunlight, shadow, and Earth's spin cause the cycle of day and night.

Teams of four students do an activity called *Spinning Globes*. Each team has a globe with colored dots affixed to four locations along the equator. Each student pretends to live in one of these locations. Teams spin the globe until the teacher says to stop. Students determine whether it is noon, midnight, or some time in-between at each of their locations.

Spinning Earth	Estimated Time
Photo of the Earth: reviewing day and night	15 minutes
Spinning globes activity	45 minutes
TOTAL	**60 minutes**

What You Need

For the class
- ❑ 1 sheet of white paper to make a color key
- ❑ 4 sticky dots, each a different bright color for color key (easy to see from a distance)
- ❑ 1 roll masking tape or clear tape
- ❑ 1 black permanent marker
- ❑ overhead transparency from the transparency packet or CD–ROM file of picture of Earth taken by *Apollo 11*
- ❑ overhead projector or computer with large screen monitor/LCD projector
- ❑ *optional:* other photographs of the Earth from space

For each group of four students
- ❑ 1 opaque Earth globe*
- ❑ a roll of masking tape or bowl to serve as a stand for the globe
- ❑ 4 sticky dots, in the same four colors as the color key
- ❑ 1 paper clip to hold dots

Note: Transparent Earth globes do not work as well as solid-color ones with this activity. Globes without built-in stands are preferable.

TEACHER CONSIDERATIONS

Unit Goals

The Earth moves with regular and predictable motion.

The Earth spins (rotates) and orbits the Sun (revolves).

The spinning of the Earth causes the apparent daily movement of the Sun and stars.

Light from the Sun shining on the spinning Earth causes day and night.

The Sun, Earth and Moon form a system.

Getting Ready

1. Prepare the Room.

   a. Set up the overhead projector or CD-ROM and monitor or projector.

   b. Set up the light bulb in the middle of the room, as in Session 3.2. Prepare to darken the room.

   c. Decide if you will rearrange the classroom. Ideally, for the Spinning Globes activity, teams of four students will be seated at tables or desks arranged in a circle around the light bulb, so that one team doesn't block the light from another.

2. Prepare the Spinning Globes materials.

   a. Globes. Inflate the Earth globes. Attach a piece of tape near the North Pole on each globe. With a permanent pen, draw an arrow on the tape indicating the counterclockwise direction as seen from above the North Pole of the Earth's spin. The tape can be removed at the end of the session.

   b. Colored sticky dots:

   • Sort colored dots for teams. Cut apart the dots, and clip together a set of four different colored dots for each team of four students.
   • Make a color key for four locations:
     i. List the four locations in large print on a piece of white paper: Hawaii, Florida, Thailand, and Egypt.
     ii. Choose which colored dot will represent each of the four locations.
     iii. Put one of each colored dot next to each location listed on the paper.
     iv. Post the color key prominently.

3. If you will not be using the CD–ROM, make a transparency of the photograph of the Earth taken by the *Apollo 11* astronauts.

4. Arrange for the appropriate projector format (computer with large screen monitor, LCD projector, or overhead projector) to display images to the class

TEACHER CONSIDERATIONS

QUESIONNAIRE CONNECTION

This session (like Session 3.2) addresses questions 1, 2, 4, and 5 on the *Pre–Unit 3 Questionnaire*. You may want to review these questions, especially 1, 4, and 5, which show Earth's rotation from an outside perspective. This perspective is also seen in the Spinning Globes activity.

Key Vocabulary

Science and Inquiry Vocabulary

Evidence

Scientific Explanation

Model

Scale Model

Prediction

Scientist

Three–Dimensional (3-D)

Two–Dimensional (2-D)

Space Science Vocabulary

Force

Mass

Gravity

Satellite

Orbit

Diameter

Sphere

System

Rotate

Revolve

Reviewing Day and Night

1. Show the picture of Earth. Show the picture of Earth taken by *Apollo 11* astronauts, and review the reasons for day and night with a series of questions:

- Ask a student to point out where on the picture it is night and where it is day.

- Point to the day side and ask, "What is making it light on this side of the Earth?" [Sunlight.]

  Students may notice the swirling clouds covering much of the Earth. The cloudless area in the middle left of the image is North Africa.

- Ask, "In this picture, which direction do you think the sunlight is coming from?" [The sunlight is coming in from the left side of the picture. We know this because the left side of the Earth is lit.] Say that although the Sun isn't in this picture (because it is so far away), we can tell that it is off to the left.

- Point to the dark part of the Earth, and ask, "Why isn't it light here?" [The Earth is blocking the sunlight; this side is in Earth's shadow.] Tell students that light travels in a straight line and cannot "go around" objects.

- Point to the night side and ask, "Why doesn't it always stay night on this part of the Earth?" [The Earth completes one spin every 24 hours.]

2. One Earth spin takes about 24 hours. Tell them that this is one example of how the movement of objects in space influences how we mark time on Earth. It takes 24 hours for the Earth to spin once. That's what we call our day, which includes both daytime and nighttime hours. You may want to add that the Earth has been spinning like this for billions of years!

Unit Goals

The Earth moves with regular and predictable motion.

The Earth spins (rotates) and orbits the Sun (revolves).

The spinning of the Earth causes the apparent daily movement of the Sun and stars.

Light from the Sun shining on the spinning Earth causes day and night.

The Sun, Earth and Moon form a system.

TEACHER CONSIDERATIONS

TEACHING NOTES

Focusing on light and dark sides of the Earth prepares students for Moon phases: Thinking about the effects of sunlight relative to the positions of the Earth and Sun in Unit 3 prepares students to understand the causes of Moon phases in Unit 4.

Session 3.3 Transparency

Picture of the Earth taken by *Apollo 11*

Sensitivity to religious views about the age of the Earth: Some students may come from families whose religious beliefs differ from the current scientific explanations of the age of the Earth and other time-frames in space science. If this is an issue for your students, you may want to point out that, to be educated citizens, all students need to understand how scientists gather evidence and what current scientific understandings are. Emphasize the importance of respect for all views, but tell them that, as students in a science class, they need to learn about and understand the scientific models and concepts.

What Some Teachers Said

"They did really enjoy the Apollo picture. We spent a good 20 minutes talking about that."

"Boy I'm telling you they want evidence for everything now. We talked about the Earth spinning like this for billions of years and they wanted to know how we knew that."

Introducing the Spinning Globes Activity

1. **Groups of four students.** Tell the class that the next activity will help them think more about different times of day and night on different parts of the Earth. Assign students to teams of four.

2. **A globe for each group.** Say that each group will get one globe and a stand (a roll of masking tape or a bowl) to set the globe on when they are not using it.

3. **Locate Hawaii, Florida, Thailand, and Egypt on a globe.** Tell them that in this activity, each person in a group will pretend to be one of four friends who live in four different locations around the Earth. Show them the equator on the globe, and point out where each of the four places are, along and a little north of, the equator: Hawaii, Florida, Thailand, and Egypt.

Alternatively, you could have students choose their own locations on the globe. We recommend, however, that these places be not far from the equator, so that the contrast between day and night will be most noticeable.

4. **Show the key.** Point out the color key posted on the wall, showing which color dot will mark each location. Say that each student will put the appropriately colored sticky dot on their assigned location. Give some hints and reference points as necessary, such as:

- The Hawaiian islands are near the center of the Pacific Ocean.

- Florida is at the southeastern corner of the United States. It's just above Cuba.

- Thailand is south of China. It is between Cambodia, Laos, and Myanmar (Burma).

- Egypt is on the southeastern side of the Mediterranean Sea. It is between Libya and Saudi Arabia.

5. **Give materials to each group of four.** Say that when teams get their materials, each student should choose their location, find it on the globe, and stick on a dot. Then, each team should place their globe on the stand and then wait for more directions. Give each team a globe, a stand for the globe (a roll of masking tape or a bowl), and four colored dots. Caution student not to play with the globes.

Unit Goals

The Earth moves with regular and predictable motion.

The Earth spins (rotates) and orbits the Sun (revolves).

The spinning of the Earth causes the apparent daily movement of the Sun and stars.

Light from the Sun shining on the spinning Earth causes day and night.

The Sun, Earth and Moon form a system.

TEACHER CONSIDERATIONS

TEACHING NOTES

Students May Need to Move Because of Light: Depending on your room arrangement, you may need to point out that some students have to move their desks or tables a bit, so that the light on their globes won't be blocked.

Key Vocabulary

Science and Inquiry Vocabulary

Evidence

Scientific Explanation

Model

Scale Model

Prediction

Scientist

Three–Dimensional (3-D)

Two–Dimensional (2-D)

Space Science Vocabulary

Force

Mass

Gravity

Satellite

Orbit

Diameter

Sphere

System

Rotate

Revolve

Spinning Globes

1. **The direction of Earth's spin.** Regain the attention of the class. Show them where the arrow indicating the direction of spin is drawn, near the North Pole. Tell students that this is the direction that the Earth spins (counterclockwise, when viewed from above the North Pole).

2. **Earth's tilt is not accurate in this model.** Mention that the actual Earth is slightly tilted, but in this model it is fine for students to keep the North Pole pointed directly up.

3. **Which student spins the globe?** Say that in each team, students will take turns spinning the Earth. Tell them who will spin first, for example, the Florida student. You may want to have them take turns in the order that the four locations are listed on the color key.

4. **Identifying whether it is day or night at their locations.** Tell them that the Florida student will slowly spin the globe counterclockwise until you say "Stop." That student will then tell their team what time of day it is at their location: "day," "night," or "in between." Then the Florida student will ask their teammates if it is noon, midnight, or in between at their locations.

Note: Please see Providing More Experience on page 387 for suggestions on having students identify sunrise and sunset.

5. **Begin the Spinning Globes activity.** Turn on the light bulb in the middle of the room and darken the room lights. Tell the first student from each team to begin slowly spinning the globe in a counterclockwise direction. Say "Stop," and give teams time to share whether it is noon, midnight, or in between at their four locations. Repeat, stopping at least four times before you end the activity.

6. **Informal assessment.** During the activity, circulate and use the opportunity for an informal assessment of whether students understand what causes day and night.

7. **Collect the materials.** When everyone on all the teams has had at least one chance to spin the globe and discuss each person's position, turn on the lights, stop the activity, and collect the materials.

Unit Goals

The Earth moves with regular and predictable motion.

The Earth spins (rotates) and orbits the Sun (revolves).

The spinning of the Earth causes the apparent daily movement of the Sun and stars.

Light from the Sun shining on the spinning Earth causes day and night.

The Sun, Earth and Moon form a system.

TEACHER CONSIDERATIONS

TEACHING NOTES

Review the procedure by modeling the "Wrong Way": Tell students that you're going to review the procedure one more time by doing everything the "wrong way." It will be their job to figure out and tell you what you are doing wrong each step of the way. Once they point out something wrong, they then tell you what would be the "right way."

This strategy allows for a thorough, step-by-step review of the procedure, and the portrayal of common student mistakes is also entertaining for students. They love to watch the teacher doing things the "wrong way," especially if you exaggerate errors or model uncooperative behavior for comedic effect. Some teachers prefer to tell their students that they do not need to raise their hands during this exercise, but can simply call out.

PROVIDING MORE EXPERIENCE

Identify Sunrise and Sunset: Depending on your students' age and abilities, you may want to make this activity more challenging by asking the "in-between" locations to identify if it's sunrise or sunset. To help students figure out whether their "dot location" is experiencing sunrise or sunset, ask, "If you spin the globe one quarter-turn counterclockwise, will your location get lighter (which would mean that it is sunrise) or darker (which would mean that it is sunset)?"

OPTIONAL PROMPTS FOR WRITING OR DISCUSSION

You may want to have students use one or both of the prompts below for science journal writing at the end of this session or as homework. These prompts could also be used for a discussion or during a final student sharing circle.

- Describe what you think would happen on Earth if the Earth stopped spinning. (Be sure to mention to students that there would still be gravity.)

- Describe what you think it would be like to be on a planet that spins twice as fast as Earth.

Key Vocabulary

Science and Inquiry Vocabulary

Evidence

Scientific Explanation

Model

Scale Model

Prediction

Scientist

Three–Dimensional (3-D)

Two–Dimensional (2-D)

Space Science Vocabulary

Force

Mass

Gravity

Satellite

Orbit

Diameter

Sphere

System

Rotate

Revolve

Unit Goals

The Earth moves
with regular and
predictable motion.

The Earth spins
(rotates) and orbits
the Sun (revolves).

The spinning of
the Earth causes
the apparent daily
movement of the
Sun and stars.

Light from the
Sun shining on the
spinning Earth causes
day and night.

The Sun, Earth and
Moon form a system.

Rotation Of Earth Plunges Entire North American Continent Into Darkness

NEW YORK—Millions of eyewitnesses watched in stunned horror Tuesday as light emptied from the sky, plunging the U.S. and neighboring countries into darkness. As the hours progressed, conditions only worsened.

At approximately 4:20 p.m. EST, the Sun began to lower from its position in the sky in a westward direction, eventually disappearing below the horizon. Reports of this global emergency continued to file in from across the continent until 5:46 p.m. PST, when the entire North American mainland was officially declared dark.

As the phenomenon hit New York, millions of motorists were forced to use their headlights to navigate through the blackness. Highways flooded with commuters who had left work to hurry home to their families. Traffic was bottlenecked for more than two hours in many major metropolitan areas.

Across the country, buses and trains are operating on limited schedules and will cease operation shortly after 12 a.m. EST, leaving hundreds of thousands of commuters in outlying areas effectively stranded in their homes.

Despite the high potential for danger and decreased visibility, scientists say they are unable to do anything to restore light to the continent at this time.

"Vast gravitational forces have rotated the planet Earth on an axis drawn through its north and south poles," said Dr. Elena Bilkins of the National Weather Service. "The Earth is in actuality spinning uncontrollably through space."

Bilkins urged citizens to remain calm, explaining that the Earth's rotation is "utterly beyond human control."

"The only thing a sensible person can do is wait it out," she said.

TEACHER CONSIDERATIONS

ASSESSMENT OPPORTUNITY

Critical Juncture—Spinning Earth: The Mount Nose activity gave students a view of the apparent movement of the Sun, seen from the perspective of a point on each student's spinning "Earth" (head). Using the model, they could see that the apparent motion of the Sun is actually caused by the spinning of the Earth. In the Spinning Globes activity, students see the spinning Earth from an outside perspective.

Quick Check for Understanding: As you watch and listen to students during this activity, notice if they understand that it is the spinning of the Earth that makes it seem to someone at a particular dot's location on Earth that the Sun moves in the sky. If not, consider doing one or more of the *Providing More Experience* activities described below.

PROVIDING MORE EXPERIENCE

1. Looking at night and day on other Earth images. Show other images of Earth taken from space. With each one, ask, "Where is it day and where is it night?" "Judging by the direction of light and the position of shadows on Earth, which direction is the Sun?" You can also do the same with images of other planets.

2. Mount Nose model using the real Sun. Do the Mount Nose model outside, using their heads to represent Earth, as before, but using the actual Sun instead of a light bulb.

3. Read humorous article, "Rotation of Earth Plunges Entire North American Continent into Darkness," (see page 388). Read the article about night falling on North America, treated as though it were a calamitous event. It is an enjoyable way to reinforce the idea that night and day are caused by the spin of the Earth. The article begins:

> NEW YORK—Millions of eyewitnesses watched in stunned horror Tuesday as light emptied from the sky, plunging the U.S. and neighboring countries into darkness. As the hours progressed, conditions only worsened.

Optional Assessment Activity: After reading the article, assign your students to write a mock letter to the editor, explaining to the author of the article why the story is ludicrous. Their letters should include the scientific explanation for the darkening event described in the article.

Key Vocabulary

Science and Inquiry Vocabulary

Evidence

Scientific Explanation

Model

Scale Model

Prediction

Scientist

Three–Dimensional (3-D)

Two–Dimensional (2-D)

Space Science Vocabulary

Force

Mass

Gravity

Satellite

Orbit

Diameter

Sphere

System

Rotate

Revolve

SESSION 3.4
Earth in Orbit

Overview

In this session, the Mount Nose model is used again, but as a demonstration that uses only one student to model Earth's orbit around the Sun, along with the Earth's spin. Students learn that it takes one year for Earth to orbit the Sun and there are 365 days in a year. The terms *rotation* and *revolution* are introduced.

Next, the class learns of the Ptolemaic model, which describes the Sun as orbiting around the Earth. Through a reading, they learn of the Copernican model, with the Earth orbiting the Sun, and how Galileo found evidence that supported this theory, which is the model used in modern science.

Students take the *Post–Unit 3 Questionnaire,* which allows them to demonstrate how their ideas about Earth's motion have changed.

3.4 Earth in Orbit	Estimated Time
Modeling Earth's orbit during a year	15 minutes
Posting key concepts	10 minutes
Reading and discussing *Copernicus and Galileo*	20 minutes
Taking the *Post–Unit 3 Questionnaire*	15 minutes
TOTAL	**60 minutes**

What You Need

For the class
- ❏ 1 lamp, light bulb and taped-down extension cord from previous session
- ❏ sentence strips for 3 key concepts
- ❏ wide-tip felt pen

For each student
- ❏ 1 copy of the reading *Copernicus and Galileo* from the student sheet packet
- ❏ 1 copy of the *Post–Unit 3 Questionnaire*

Unit Goals

The Earth moves with regular and predictable motion.

The Earth spins (rotates) and orbits the Sun (revolves).

The spinning of the Earth causes the apparent daily movement of the Sun and stars.

Light from the Sun shining on the spinning Earth causes day and night.

The Sun, Earth and Moon form a system.

THE KEY CONCEPT WALL

WHAT WE HAVE LEARNED ABOUT EVIDENCE AND MODELS

1. Evidence is information, such as measurements or observations, that is used to help explain things.

2. Scientists base their explanations on evidence.

3. Scientists question, discuss, and check each other's evidence and explanations.

4. Scientists use models to help understand and explain how things work.

5. Space scientists use models to study things that are very big or far away.

6. Models help us make and test predictions.

Note: These three concepts are not introduced in Units 2, 3, and 4. However, if you introduced them in Unit 1, keep them on the concept wall:

7. Every model is inaccurate in some way.

8. Models can be 3-dimensional or 2-dimensional.

9. A model can be an explanation in your mind.

WHAT WE HAVE LEARNED ABOUT SPACE SCIENCE

Unit 3: How does the Earth move?

3.1

The Sun and stars appear to move across the sky.

3.2

The Earth spins.

The spinning of the Earth makes it look like the Sun and stars are moving.

It takes the Earth 24 hours to spin once.

The Earth spinning in sunlight causes day and night.

3.4

The Earth orbits the Sun.

It takes Earth a year to orbit the Sun once.

The Sun, Earth, and Moon form a system.

Getting Ready

1. Prepare the Room.

  a. Set up the light bulb for the demonstration. It needs to be in an area where one student can walk around it to model Earth's orbital and rotational movements while the class observes.

  b. Prepare to darken the room.

2. Make photocopies. Decide if you will use one or both pages of the *Copernicus and Galileo* reading. Make a copy of the reading and the *Post–Unit 3 Questionnaire* for each student.

3. Write the following three key concepts on sentence strips and have them ready to post on the concept wall during the session.

The Earth orbits the Sun.
It takes Earth a year to orbit the Sun once.
The Sun, Earth, and Moon form a system.

Modeling the Earth's Movement During One Year

In this demonstration, the class remains seated as one student demonstrates the Earth orbiting the Sun.

1. Briefly review how the Earth spins. With the light bulb on and the room darkened, hold up a globe. Point to a place on the night side, and ask someone to summarize what they have learned about the Earth's movement through the Mount Nose model and the Spinning Globes activity.

2. Length of a day. Ask how long it takes for the Earth to make one complete spin. [24 hours.]

3. A student models one day. As a review, have a student volunteer stand by the light bulb and spin around once, as with the Mount Nose model. Review, "For a person on Mount Nose, why does the Sun seem to move in the sky?" [The Earth spins.] Remind them that the stars seem to move for the same reason.

Unit Goals

The Earth moves with regular and predictable motion.

The Earth spins (rotates) and orbits the Sun (revolves).

The spinning of the Earth causes the apparent daily movement of the Sun and stars.

Light from the Sun shining on the spinning Earth causes day and night.

The Sun, Earth and Moon form a system.

TEACHER CONSIDERATIONS

TEACHING NOTES

Reading level: The reading level of page 1 of the reading is appropriate for most third and fourth graders. Page 2 is for students who are interested in further information on the topic, and who are able to read at a slightly higher level.

Key Vocabulary

Science and Inquiry Vocabulary

Evidence

Scientific Explanation

Model

Scale Model

Prediction

Scientist

Three–Dimensional (3-D)

Two–Dimensional (2-D)

Space Science Vocabulary

Force

Mass

Gravity

Satellite

Orbit

Diameter

Sphere

System

Rotate

Revolve

Unit Goals

The Earth moves with regular and predictable motion.

The Earth spins (rotates) and orbits the Sun (revolves).

The spinning of the Earth causes the apparent daily movement of the Sun and stars.

Light from the Sun shining on the spinning Earth causes day and night.

The Sun, Earth and Moon form a system.

4. Length of a year. Ask if anyone knows another movement that the Earth makes besides spinning. [The Earth completes an orbit around the Sun once a year.] Explain that the word *orbit* means "to go around in a circular or oval path." Have the volunteer walk once around the light bulb. Emphasize that even though many ancient peoples thought that the Sun went around the Earth, we now know that the Earth actually orbits the Sun. Tell students that people argued about this model for a long time, but scientists have gathered lots of evidence for the Sun-centered model through observations, measurements, and mathematics.

5. Volunteer models both spin and orbit. Ask, "How many days are in a year?" [365. The Earth spins 365 times during one full orbit of the Sun.] Ask your volunteer to slowly (and carefully!) spin and simultaneously "orbit" partway around the "Sun" with each spin. Stop the student after a few spins, before he or she gets too dizzy. Say that the Earth spins 365 times every time it orbits the Sun once.

6. Rotate/Revolve. Introduce the words that scientists use for spinning and orbiting: to *rotate* means "to spin" and to *revolve* means "to orbit." The Earth rotates once a day and revolves once a year.

7. Space objects moving and marking time. Make the connection between space science and daily life by pointing out that the way objects move in space influences how we mark time on Earth. One Earth orbit is what we call one year.

8. Lunar orbit around Earth takes about a month. Ask, "What sky object goes around in its orbit about once a month?" [The Moon orbits (revolves around) the Earth every 29.5 days.]

Earth's moon is the only large object that orbits the Earth.

9. A system of objects in predictable motion. Say that a system is a group of objects that form a whole, and move or work together. The Earth, Moon, and Sun form a system. All three objects move in a regular, predictable way.

TEACHER CONSIDERATIONS

TEACHING NOTES

Different Constellations can be seen throughout the Earth's orbit:
As Earth orbits the Sun throughout a year, we are able to see different constellations, depending on what side of the Sun we are on.

QUESTIONNAIRE CONNECTION

After the volunteer models how the Earth moves relative to the Sun, you may want to revisit question #3 on the *Pre–Unit 3 Questionnaire*.

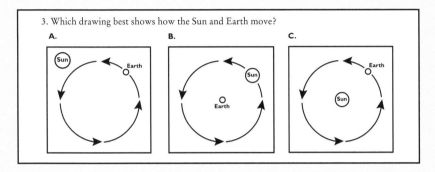

SCIENCE NOTES

The Moon does rotate: It's probably better not to discuss this with students now, but the Moon does rotate, although its spin is very slow. The Moon spins on its axis one time during every orbit around the Earth.

Because of the pull of gravity between the Earth and Moon, one side of the Moon is "locked," facing in the direction of the Earth. As the Moon orbits, that side always faces Earth. Imagine that you were looking at the Earth–Moon system from an outside perspective. Picture a big X on the side of the Moon facing Earth. As the Moon orbits the Earth, the X faces different directions in space, although it keeps one side always facing Earth. (This is much easier to visualize if you use a model. Use two spheres and put an X on the sphere representing the Moon.)

Key Vocabulary

Science and Inquiry Vocabulary

Evidence

Scientific Explanation

Model

Scale Model

Prediction

Scientist

Three–Dimensional (3-D)

Two–Dimensional (2-D)

Space Science Vocabulary

Force

Mass

Gravity

Satellite

Orbit

Diameter

Sphere

System

Rotate

Revolve

10. **Post key concepts.** Point out that the class has used models to explain how the Earth moves. Add the following key concepts to the concept wall:

The Earth orbits the Sun.
It takes Earth a year to orbit the Sun once.
The Sun, Earth, and Moon form a system.

Reading: *Copernicus and Galileo*

1. **Read about history of ideas regarding Earth's movement.** Tell the class that they're going to read about the history of the discovery of how the Earth and Sun really move.

2. **Ptolemy.** Tell students that about 2,000 years ago, a man named Ptolemy (Toll'-emy) taught that Earth is the center of everything. He didn't think that the Earth moved at all. He explained that the Sun, stars, and planets are all orbiting the Earth. People thought that his model made sense. His model was passed on for more than a thousand years. The observations people made back then seemed to give evidence that supported Ptolemy's idea.

3. **Read *Copernicus and Galileo*.** Depending on students' reading abilities, you may want to have them do independent, paired, or shared reading. Help students with the pronunciation of the names in the title. Tell them that if they finish the reading on page 1, they should continue reading more on page 2.

Unit Goals

The Earth moves with regular and predictable motion.

The Earth spins (rotates) and orbits the Sun (revolves).

The spinning of the Earth causes the apparent daily movement of the Sun and stars.

Light from the Sun shining on the spinning Earth causes day and night.

The Sun, Earth and Moon form a system.

TEACHER CONSIDERATIONS

TEACHING NOTES

Purposes of the Reading:

- *Reviews contrasting models for Earth–Sun system.* The reading reviews the Earth-centered (Ptolemaic) model, and why it is inaccurate. This may help students who still have a similar misconception revise their thinking. The reading also reviews the Sun-centered model (Copernican), which fits all the evidence and is now accepted as fact.

- *Provides an example of how history and science can be inter-related.* The reading shows how, throughout history, ideas change as people make observations and find more scientific evidence relating to questions about the natural world. It brings up an example of an idea that encountered strong resistance.

- *Illustrates the nature of science. Scientists attempt to find the best explanation for the available evidence.* This is one of the most important components of scientific investigation.

- *Introduces important astronomy pioneers, Copernicus and Galileo.* Copernicus invented the Sun-centered model of the Solar System. Among other things, Galileo made important discoveries related to gravity, telescopes, and the Solar System.

- *Illustrates the importance of the telescope* as an evidence-gathering tool for space scientists.

More on the invention of the Telescope: Hans Lippershey was a Dutch spectacle maker. It is not certain if he was the first to invent the telescope, but he was the first to apply for a patent. He is generally credited as the inventor of the telescope, and it is his story included in the student reading. The patent was denied because the idea had become common knowledge by the time he applied for it.

The word telescope *comes from two Greek words: tele means "far" and skopein means "to look or see"," thus teleskopos means "far-seeing")*

More on the history of the Sun-centered model of the Solar System: The *Background Information* for teachers has more information on the history of how people came up with the heliocentric model of the Solar System.

Key Vocabulary

Science and Inquiry Vocabulary

Evidence

Scientific Explanation

Model

Scale Model

Prediction

Scientist

Three–Dimensional (3-D)

Two–Dimensional (2-D)

Space Science Vocabulary

Force

Mass

Gravity

Satellite

Orbit

Diameter

Sphere

System

Rotate

Revolve

COPERNICUS AND GALILEO, PAGE 1

Name:_____

Copernicus and Galileo

Long ago, most people believed that the Earth is at the center of everything. They thought that the Sun and stars orbit the Earth. In 1543, a man came up with a new explanation for the way the Sun and stars seem to move across the sky. His name was Copernicus (Co-PER-nick-us). He lived in Poland. He spent years observing the planets, Moon, stars, and Sun. He used math to try to create a model explaining what was going on. The more he observed, the more evidence he found that didn't fit the old model. He thought up a new model that fit all his evidence: he decided that the Earth must orbit the Sun. He wrote his ideas down, but he knew that his new model would upset many people. He waited until he was dying to put out a book on his ideas.

A young man named Galileo liked new ideas. He lived in Italy. He was always asking his teachers for evidence for their explanations. He spent many years studying objects in the sky, too. Galileo heard about the ideas that Copernicus had written down. He thought that they made sense.

Galileo also heard about a new tool that someone had invented—the first telescope. Galileo started making telescopes, too. In 1610, Galileo became the first person to study the night sky with a telescope. With the telescope, Galileo could see moons orbiting the planet Jupiter. This was evidence that not everything orbits Earth. Galileo shared his ideas with many people.

But some important people still thought that everything orbits Earth. They didn't like the Copernicus model. They were very angry with Galileo for agreeing with Copernicus. They arrested Galileo and made him say that he didn't believe in the Copernicus model. They made him say that he thought that everything orbits the Earth. Galileo was kept under house arrest for the rest of his life.

Today, we have telescopes much more powerful than Galileo's. We even have telescopes in space. Astronomers have been observing and recording evidence about the movements of sky objects for hundreds of years. From space, we've seen that the Earth spins. From space, we've seen that Earth and other planets orbit around the Sun. We have lots of evidence that the Copernicus model is the best explanation for the movement of the Sun, planets, and moons.

Discussing the Reading

1. Discuss the Reading. After the reading, ask the following questions, and allow as much discussion as possible:

• Why do you think that the inaccurate theory that the Sun orbits the Earth was accepted for more than a thousand years? Why do you think that many people did not like the Copernican model? [From Earth, it seems as though the Sun goes around the Earth. Also, people may have found the idea that Earth is the center of everything more comfortable.]

• What evidence did Galileo use to show that the model of the Earth orbiting the Sun is more accurate? [Using a telescope, he found evidence of moons orbiting Jupiter; this meant that not everything orbits the Earth.]

TEACHER CONSIDERATIONS

COPERNICUS AND GALILEO, PAGE 2

TEACHING NOTES

Additional evidence that supported the Copernican model:

- Using the Sun-centered model, Galileo and Copernicus were able to predict the calendar better than those using the Earth-centered model.

- Galileo tracked spots on the Sun, and could see that the Sun is spinning.

- The Earth-centered model includes crystal spheres around the Earth. Each planet has its own crystal sphere. Comets have been seen coming from far away and moving closer to the Sun. They couldn't have done that if there were crystal spheres.

- By studying Venus with a telescope, Galileo was able to see the phases of Venus. These phases only make sense with a Sun-centered model of the Solar System. (Students will learn more about what causes phases in Unit 4.)

Name: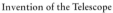

Galileo, the Father of Modern Science

Galileo made many discoveries that changed how we see the world. He taught that explanations should be based on all the best evidence. Because of Galileo's writing and his many discoveries, he is sometimes called "the father of modern science." Here are some of the things Galileo did:

- observed objects in the sky with a telescope before anyone else did
- made telescopes work better
- discovered mountains on the Moon
- discovered the moons of Jupiter
- discovered Saturn's rings
- discovered the phases of Venus
- discovered sunspots and used their movement as evidence that the Sun spins
- helped invent the thermometer
- made discoveries about pendulums
- made discoveries about how objects fall

Invention of the Telescope

During Galileo's time, many people wore eyeglasses so they could see better. Like glasses today, the eyeglasses had two lenses, one for each eye. The lenses were made of curved glass. One day, a man who made glasses let some children play with two lenses. They took the two lenses, held them apart, and looked through them. The children were amazed that faraway things looked bigger and closer through the lenses. The glasses maker tried it, and he was also amazed. He put the two lenses at either end of a tube. This was the first telescope. Galileo learned about the invention and started making one himself. Galileo did not invent the telescope, but he made it better. He was the first person to look at objects in the night sky through a telescope.

Galileo and the Church

In Galileo's time in Italy, if people didn't agree with the teachings of the Catholic Church, they could be arrested and even executed. The Church taught that the Sun, stars, and planets all orbit the Earth. They taught that the Earth does not spin. Because Galileo taught the Copernicus model, Church leaders arrested him and put him on trial. They made Galileo say that he believed the Sun, stars, and planets orbit the Earth. They made him say that he did not believe in the Copernicus model. They did not allow Galileo to leave his home for the rest of his life. In modern times, Catholic leaders have said that the Church made mistakes in how they treated Galileo and when they tried to stop people from supporting the Copernicus model.

We now know that the Copernicus model is accurate. We now know that the Earth and other planets in the Solar System orbit the Sun. We know there are planets orbiting other stars.

POST–UNIT 3 QUESTIONNAIRE, PAGE 1

B.Session 3.4 Student Sheet   Name _____

Post–Unit 3 Questionnaire, Page 1

Note: Pictures are not to scale

1.Which drawing best shows how the Sun and Earth move?

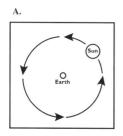 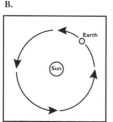 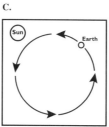

A.   **B.**   **C.**

2. Which picture shows what causes day and night? Circle A, B, or C.

A.Earth Orbiting the Sun   **B. Sun Orbiting the Earth**   **C. Earth Spinning**

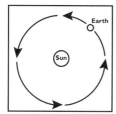 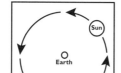 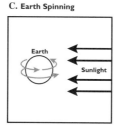

(Over)

The Post–Unit 3 Questionnaire: How the Earth Moves

1. **Find out how their ideas may have changed.** Tell the class that they have used models to study the movements of the Earth since they took the *Pre–Unit 3 Questionnaire*. Let them know that they will now get a chance to take it again to see how their ideas may have changed. Remind them that the ability to change one's ideas based on evidence is a sign of being a scientist.

2. **Review the questionnaire guidelines.** Mention that the questions are the same, but the order of the questions and answers in the questionnaire has been changed.

3. **Emphasize the importance of working independently.** Say that the questionnaire is designed to find out what each student is thinking. If they don't know an answer, they should respond the best they can.

4. **Distribute pencils and questionnaires.** Let students know what to do if they finish early, and give everyone adequate time to finish before collecting the questionnaires.

TEACHER CONSIDERATIONS

OPTIONAL PROMPTS FOR WRITING OR DISCUSSION

You may want to have students use one or more of the prompts below for science journal writing at the end of this session or as homework. These could also be used for a discussion or during a final student sharing circle.

- Describe how the Earth has moved in relation to the Sun in the last whole year of your life.

- How many times have you orbited the Sun in your whole life? How many spins has the Earth made since the beginning of the month you are in now?

- When it was a new idea, some people refused to accept the Copernicus model of the Earth orbiting the Sun. Why do you think people did not accept this new idea?

Name:_____

Post-unit 3 Questionnaire continued

3. Here is an enlarged person standing on the Earth. It is noon for this person. Which direction is the Sun? Circle A, B, C, or D.

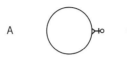

4. Why does it look like the Sun is moving down in the sky at sunset? Circle the best answer.

A. The Earth goes around the Sun.
B. The Sun goes around the Earth.
C. The Sun is spinning.
D. The Earth is spinning.

The Sun at 5:55 p.m.

The Sun at 6:00 p.m.

5. Where is it night on Earth in this picture? Color in the areas where it is night.

Earth

Sunlight

PROVIDING MORE EXPERIENCE

Your Age on Other Planets: The Earth takes about 365 Earth days to orbit the Sun. Some planets take less time, and some take more. So if a person who is eleven years old on Earth had spent their whole life on Mercury, which takes much less time to orbit the Sun, they would be many Mercury years older. If they had spent their life on Jupiter, which takes much longer to orbit the Sun, they would be many fewer Jupiter years old. At the following website, students can type in their birth date and calculate how many years old they would be on other planets:
http://www.exploratorium.edu/ronh/age/index.html

What Some Teachers Said

"It was very informative to see the extent of knowledge before and after the lessons. I was extremely pleased with the outcome of the post questionnaire because it indicated that the students understood the concepts!"

"My students had little or no knowledge of these concepts and many misconceptions were cleared up by the end."